What's for lunch?

Peanuts

Library of Congress Cataloging-in-Publication Data
Llewellyn, Claire.
   Peanuts / Claire Llewellyn.
   p.   cm. -- (What's for lunch?)
   Includes index.
   Summary: Presents facts about the peanut, including where and how it is grown,
   harvested, and marketed and what other products are made from peanuts.
   ISBN 0-516-20839-X
   1. Peanuts --Juvenile literature. 2. Peanut products--Juvenile
literature. [1. Peanuts. 2. Peanut products] I. Title. II. Series.
Llewellyn, Claire. What's for Lunch?
SB351. P3L63 1998   97-34925
641.3'56596--dc21   CIP
                                            AC

96 Leonard Street
London
EC2A 4RH

First American edition 1998 by
Franklin Watts
A Division of Grolier Publishing
Sherman Turnpike
Danbury, CT 06816

ISBN   0-516-20839-X (lib. bdg.)
ISBN   0-516-26222-X (pbk)

Editor: Samantha Armstrong
Series Designer: Kirstie Billingham
Consultant: American Peanut Council
Reading Consultant: Prue Goodwin, Reading and Language
Information Centre, Reading.

Printed in Hong Kong

What's for lunch?

Peanuts

Claire Llewellyn

CHILDREN'S PRESS®

A Division of Grolier Publishing
NEW YORK • LONDON • HONG KONG • SYDNEY
DANBURY, CONNECTICUT

Today we are having peanut butter for lunch.
Peanut butter is made from peanuts.
Peanuts contain **protein**, **vitamins**,
minerals, and **fiber**.
They give us **energy** and help us grow
and stay healthy.

A nut is a hard, dry fruit.

Inside every nut is a **seed**.

Hazelnuts, acorns, and chestnuts
are all different kinds of nuts.

Peanuts are not nuts.

They are the seeds of the peanut plant.

The seeds grow in **pods** under the ground.

Peanuts grow in warm countries
around the world.
In spring, peanut farmers
sow peanut seeds in their fields.
The seeds are peanuts from
last year's **crop**.
The seeds **sprout** quickly and
grow into small plants.

The plants grow yellow flowers
on long stalks called **pegs**.
Instead of growing up toward the sun,
the pegs grow down toward the earth.
They push their way through the soil.
Under the ground the flowers change
into pods, and the peanuts start to grow.

After about five months,
the peanuts are ready to **harvest**.
A machine digs up the plants,
turns them over, and leaves the pods
to dry in the sun.

A few days later, a **combine harvester**
pulls the pods off the plants and
piles them into a wagon.

Peanut farmers save some of the crop
for next year's seed.
The rest is packed into sacks.
The sacks are delivered to factories
all over the world.

In the factories, the peanuts are checked. Any peanuts that will not be good to eat are removed.

Sometimes peanuts are **roasted**
inside their shells.
Or they are taken out of their shells
and salted, or coated with honey.

When they are ready to eat
the peanuts are packed into bags.
The bags are sealed to keep the peanuts fresh.
They are now delivered to stores
and supermarkets.

Peanuts can be made into peanut butter.
To do this the peanuts are **ground** and
sugar and salt are added to the **paste**.
This is beaten until it is very runny and
can be poured into jars.
It **sets** into peanut butter.

Oil made from peanuts, called
peanut oil, is used in cooking.
It is especially good for Chinese **stir-fry** cooking
because the oil does not smoke,
even when it is very hot.
Peanut oil is also added to hand cream,
lipstick, and paint.

We eat peanuts in all sorts of ways.
Peanuts can be added to cookies,
stirred into different dishes,
or just eaten on their own.
They are crunchy and tasty and
make a very healthy snack.

Glossary

combine harvester   a large machine that removes the peanut pods from the plants

crop   what farmers grow in their fields

energy   the strength to work and play

fiber   something found in food that helps us to digest properly

ground   crushed into small bits

harvest   gather the crop from the fields

mineral   a material that is found in rocks and also in our food. Minerals are important to keep us healthy

paste   what is left when peanuts are ground

peg   the stalk of the peanut plant

pod	the shell that the peanuts grow inside
protein	something found in food, such as peanuts, that helps keep us healthy
roast	to cook in an oven
seed	the part of the plant from which a new plant grows
set	when something runny becomes solid
sow	to plant seeds in a field to grow a crop
sprout	to grow shoots
stir-fry	a special way of cooking food in a Chinese wok or a frying pan
vitamin	something found in food, especially milk, fruits, and vegetables, that keeps us healthy

Index

Picture credits: Bruce Coleman 11 (John Murray); Courtesy of Percy Dalton Famous Peanut Company Ltd. (From Cache Collection) 19, 23; F. Duerr and Sons Ltd. 24; FLPA 7 (Silvestris), 8 (L. Wiame/Sunset); Grant Heilman Photography Inc. 6 (Jane Grushow), 10 (Runk/Schoenberger), 12 (Grant Heilman), 14-15 (Arthur C. Smith III); Holt Studios International 8, 13 (Nigel Cattlin); Panos Pictures 16 (Tina Gue), 17 (Jeremy Hartley); Robert Harding Picture Library 26; Zefa 20 (Bramaz); Steve Shott cover; All other photographs Tim Ridley, Wells Street Studios, London.
With thanks to Thomas Ong and Redmond Carney.